Introduction and History

Clare College is one of the most ancient and venerable of Cambridge's colleges. It is the second oldest college in the university, after Peterhouse, and the oldest college on the Backs – that pleasant area of the city that takes its name from the backs of the colleges that lead down to the River Cam. Clare Bridge, which links Old Court with the College gardens on the far side of the Cam and Clare's newer courts beyond Queens Road, is the oldest bridge over the river in Cambridge, begun in 1638; and the wall that divides the Scholars' Garden from King's College lawn is a medieval survival.

The College can also be hailed as the pioneer of the collegiate model: when it was founded in 1326 it was the first educational institution in Britain to provide for students and teachers living and working together in a single community, and was the template for all subsequent Oxbridge colleges and for the much later residential communities at Harvard and Yale. Nearly seven centuries on, this collegiate model is key to the special character and strength of a Cambridge education.

Clare takes its name from its foundress, Lady Elizabeth de Clare, a granddaughter of King Edward I who became Lady de Burgh when she married. She rescued the College

Right: Elizabeth de Clare.

in 1336 with a generous endowment when it was struggling to survive after a disastrous fire, just over a decade after Richard de Badew had set it up as University Hall.

Lady Elizabeth was married three times to influential noblemen, but was also wealthy in her own right after her only brother, Gilbert de Clare, was killed at the Battle of Bannockburn without leaving a son. She and her two sisters therefore inherited their father's substantial wealth, and all three of them were bound up in the tempestuous politics of the time, with their status as heiresses making them valuable properties to be traded in the marriage marketplace. But after the death of her third husband in 1322, when Elizabeth was probably aged around thirty, she did not remarry, but used her riches and her lengthy widowhood (she died in 1360) on good works. The College was one of the major beneficiaries, both during her life and in her will; already by 1339 it was known as Clare Hall, and she provided it with its statutes in 1359. Her portrait now takes pride of place behind High Table in the Hall.

The original endowment provided for the maintenance of a maximum of fifteen 'scholars' (subsequently called 'fellows'), of whom no more than six were to be bound by priestly orders. Provision was also made for ten 'poor scholars' who were to be maintained by the College up to the age of twenty. Lady Elizabeth's statutes showed an enlightened attitude to education at the time, offering learning as a path to advancement and stating that 'the knowledge of letters... sendeth forth its students, who have tasted of its sweetness, fit and proper members in God's Church and the State, to rise to diverse heights, according to the claim of their deserts'.

Little more is known of Clare's early years, since a fire in 1521 destroyed most of the early records as well as the College's medieval buildings. Their replacements were makeshift, so by the early seventeenth century, when Clare had grown considerably in both wealth and numbers, there was an urgent need both to expand and to provide more suitable accommodation. The College's

magnificent Old Court and its bridge over the Cam were the fruits of this expansion; and the majesty of the buildings was matched by the distinction of the College's fellowship during these centuries. As well as many scholars eminent in their fields, including Isaac Newton's successor as Lucasian Professor of Mathematics in the university, the College educated future Bishops of Ely and London, an Archbishop of Canterbury, a Chancellor of the Exchequer and a Prime Minister; it has to be admitted, however, that the output of Clare's only Poet Laureate, William Whitehead, has not stood the test of time.

During the nineteenth century – when the word 'Hall' was changed to 'College' in its name – Clare languished somewhat in the academic stakes, despite boasting a Master, Edward Atkinson,

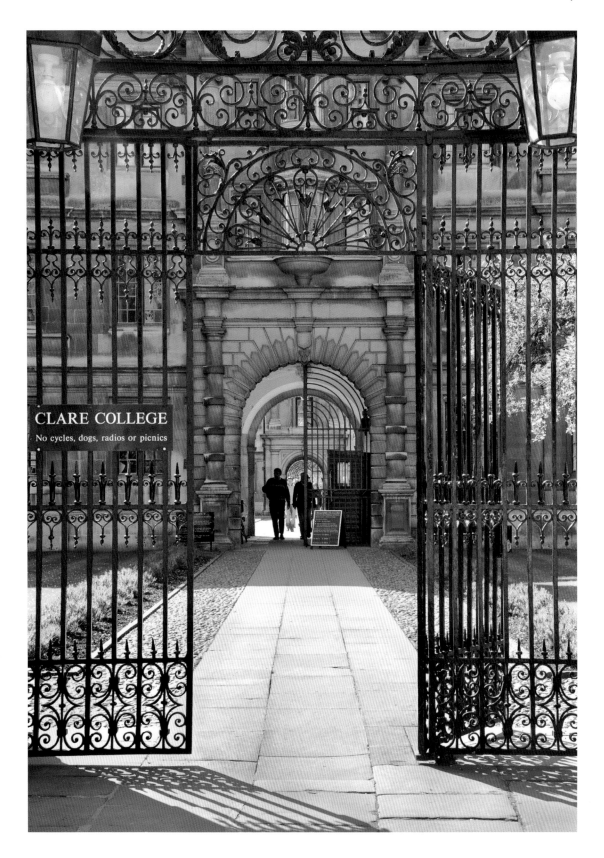

MUSIC AT CLARE

Clare College's reputation for musical excellence and its many remarkable contributions to music over recent decades – choral, instrumental and academic – have resulted in it admitting more undergraduates to read for a degree in music than any other college in the University of Cambridge. Under its highly distinguished Directors of Music, first John Rutter and then, since 1979, Timothy Brown, the College has been dedicated to nurturing young musicians in the early stages of their careers by giving them many opportunities to perform and by encouraging and supporting them in the development of their talents, whether they are reading music or combining a love for music with studying another subject.

Clare has launched many professional musicians onto the world stage. John Rutter was encouraged while an undergraduate at Clare to devote himself to composing, and his music is now part of the core repertoire of choral groups the world over. As an undergraduate choral scholar, Elin Manahan Thomas developed as a soloist and made headline news as the first singer for almost

three centuries to perform Bach's seductive aria 'Alles mit Gott'. The conductor Sir Roger Norrington and the musical entertainer Richard Stilgoe are former choral scholars, while former organ scholars include the organist David Dunnett and the conductors Ivor Bolton and Richard Egarr.

Instrumentalists too have flourished. Two Clare pianists have been winners in young musician competitions; the flautist Daniel Pailthorpe became Principal Flute of the English National Opera Orchestra at the age of twenty-four and is currently Co-Principal with the BBC Symphony Orchestra; and several violinists and cellists are leaders or members of professional ensembles, including Andrew Manze who is Artistic Director of The English Concert.

Clare Choir, established in 1866 and becoming a mixed choir in 1971, has an enviable national and international reputation. Beyond its primary task of leading chapel services, the Choir gives frequent concerts in the UK and abroad and receives regular invitations to perform with many of Europe's

leading orchestras and conductors. In 2000 it became the first Oxbridge mixed choir to perform at the BBC Promenade Concerts, singing Bach's *St John Passion*. In 2002 the Choir took part in a sell-out tour to Europe and the USA, performing Handel's *Jephtha* with the Orchestra of the Age of Enlightenment. In 2005 its national importance was reflected when it was invited to perform at Britain's National Holocaust Memorial Day commemoration in Westminster Hall. And in 2008 the Choir commissioned and gave the world premiere performance of a major new work by Sir John Tavener, *Ex Maria Virgine*.

Given this international standing, Clare Choir's amateur status comes as a surprise to many. The choral scholars and volunteers who make up the Choir combine their singing with academic study across the full range of subjects, from English to medicine; and despite the dedication required by singing at world-class level, many choristers also find time for other extra-curricular activities, ranging from sport to poetry and drama. Moreover, the discipline involved in undertaking professional engagements of a high order and combining them successfully with study makes the Choir an invaluable training ground for musicians, and prepares them for coping with the professional world beyond university. Many former members of Clare Choir are now distinguished professional soloists, including the bass Simon Bailey, the soprano Ruth Holton and the tenor Nicholas Mulroy. Others sing in leading vocal ensembles, such as the Monteverdi Choir, The Sixteen and the Tallis Scholars.

Top: The West Range from Clare Bridge.

Above: The Avenue from the Queens Road entrance.

whose tenure of his post (1856–1915) is one of the longest on record. But numbers and reputation grew during the twentieth century, to the point where further expansion became necessary: Memorial Court and Thirkill Court were built beyond Queens Road on the west bank of the Cam; the outpost of the College within the city which has become known as the Colony was developed on property Clare has owned since the Middle Ages near Castle Mound; and at the beginning of the twenty-first century Lerner Court has been constructed in an area next to Memorial Court which used to house bicycle sheds and outhouses.

Opposite: Aerial view of Clare College, King's Chapel and the centre of Cambridge.

The pioneering spirit that has been a feature of Clare since its first statutes set out its ethos in 1359 has continued to influence its progress in the modern world. In the 1960s the College was one of the first in Cambridge to provide facilities specifically for graduate students when it set up

EXTRACT FROM 'THE LAST MEETING' BY SIEGFRIED SASSOON, FLIXÉCOURT, MAY 1916

My heart is fooled with fancies, being wise;
For fancy is the gleaming of wet flowers
When the hid sun looks forth with golden stare.
Thus, when I find new loveliness to praise,
And things long-known shine out in sudden grace,
Then will I think: 'He moves before me now.'
So he will never come but in delight,
And, as it was in life, his name shall be
Wonder awaking in a summer dawn,
And youth, that dying, touched my lips to song.

Clare Hall on land beside Herschel Road; and in 1972, even more momentously, Clare became one of the first three Cambridge men's colleges to admit undergraduate women, a change which had a dramatic effect on the nature of the College, particularly in its immediately enhanced standing in examination results. Clare had appointed its first women Fellows a year earlier, in 1971.

Recent Clare Fellows and alumni have reached the forefront of academic distinction, and include such glittering figures as Sir Geoffrey Elton, the eminent historian; Nobel laureate James Watson, co-discoverer of the structure of DNA; and Sir Andrew Wiles, the mathematician who finally solved Fermat's Last Theorem. Siegfried Sassoon, the poet who chronicled the horrors of the Great War, was at Clare, as was Sir David Attenborough, the prominent broadcaster and naturalist.

Above: David Attenborough as an undergraduate.
Left: Siegfried Sassoon, by Sir William Rothenstein.

The Old Buildings

Clare College, tucked away from the bustle of Cambridge's city centre down Senate House Passage and along Trinity Lane, forms part of the superb architectural grouping of King's College Chapel, the Old Schools and its own east and south fronts. Its façade is set back from the road, the result of somewhat acrimonious negotiations with King's in the early seventeenth century when Clare wanted both to expand and to acquire land on the west bank of the Cam, then owned by King's; in return for this land, Clare agreed to cede part of its property here, and to set its entrance back from Trinity Lane. The small entrance forecourt is fronted by gateposts carved in 1675 and iron gates dating to 1713–15. Made by an ironsmith called Warren they, in common with his other gates for Clare by the bridge and at the exit from the Backs to Queens Road, are among the most distinguished examples of early eighteenth-century decorative ironwork in existence.

Left: The entrance court and Chapel.

Below: The front gates with King's Chapel in the background.

Old Court

Old Court is unique among the Cambridge colleges founded before 1800 in having maintained its architectural integrity and its resistance to later accretions; apart from an early nineteenth-century face-lift which enlarged the windows and introduced a few touches of Regency taste, it has remained truc to its seventeenth-century self.

Its predecessor was a medieval court half the size, which was both too small for a growing community and in a parlous state. In common, therefore, with other Cambridge colleges in the early seventeenth century which were being rebuilt, faced in stone and generally upgraded, the Fellows of Clare decided to demolish their old court in stages and replace it with a building both large enough for the College's needs and worthy of its status. The work was much interrupted, particularly by the Civil War, but was almost complete by the turn of the century. Thanks to the bursar Barnabas Oley, who kept detailed accounts, the building is very well documented.

The walls are of brick faced with stone on seven of its eight façades; money was saved by leaving in

Left: The West and South Ranges from the bank of the Cam.

Below: The West Range and the Scholars' Garden.

unfaced brick the inaccessible back wall of the North Range, facing Trinity Hall. The East Range was the first to be built, with its foundation stone laid in 1638, although the College had begun to stockpile material as early as 1635. The building of the bridge over the Cam went along in tandem, and both were completed by 1641. The South Range was next, taking just over two years to build and finished by the end of 1642. Work on the West Range had meanwhile started but was interrupted when Oliver Cromwell confiscated the materials intended for it because he needed them to strengthen Cambridge Castle. There was a long gap during the war and the interregnum, and work did not resume until 1662. It then proceeded sporadically, along with the building of the North Range, with its formal rooms; the Hall was opened in 1693, but it was to be another twenty years before the whole of the court was complete, when the Master's Lodge in the north-west corner was finished in 1715. The total cost was £15,478 10s 2¼d.

LADY ELIZABETH'S BELL

Lady Elizabeth's bell, now hanging outside B staircase opposite the Porters' Lodge in the eastern gateway to Old Court, is a recent acquisition, though it is an authentic fourteenth-century bell and its inscription indicates that it once had a connection with Clare's foundress. It was by chance that the bell's modern donor, Mary Bliss, discovered it in 2002 in an auction sale near Gloucester, listed merely as 'one seventeen-inch bell' and catalogued among a long list of garden implements; she paid £52 for it. The inscription reads:

ELIZABETIDEBVRGOIES VS CRISTVS, and expert opinion has confirmed its early date and that it was originally donated by Elizabeth de Burgh, or Lady Clare. Having found out about its history, Mary Bliss dispensed with the nineteenth-century clapper inside, and substituted it with a medieval clapper which a bellhanger just happened to have in his workshop. She then donated the bell to the College and was present at a short dedication ceremony also attended by the Master, the Dean and a senior Fellow, Dr Gordon Wright.

Old Court can be said to encapsulate the history of English architecture during the seventeenth century, from the last gasps of Gothic betrayed by the fan vaulting in the east gateway, via the Palladian influences exhibited by the West

Range and the Wren touches discernible in the North Range to the mature Classicism seen in the Master's Lodge. There seems to have been no overall controlling hand, and no evidence – despite wishful Clare tradition – that Inigo Jones was involved. The design appears to have been the result of a loose collaboration between the College authorities, the men entrusted with overseeing the work and the stone masons. The names of some of the key players have survived in Barnabas Oley's records. One of them, John Westley, was the building contractor who was responsible for the East and South Ranges; he was left £10 in Oley's will and spoken of as 'that good workman that built the Colledge'. Two others were the Grumbolds, father Thomas and son Robert. Thomas, a skilled stone mason, is credited with much of the stonework on the earlier ranges, and with the bridge over the Cam, while Robert's name appears in connection with the design and building of the North Range.

This range show clear signs of the influence of Christopher Wren, on whose library for Trinity College Robert Grumbold worked. Grumbold visited Wren in London several times during the building of the library, and certainly imported some of Wren's ideas into his work for Clare.

Previous pages: Old Court looking east.

Below: Old Court by night.

It is very likely that it was he who designed and built at least the North Range, since he was paid for a drawing of it in 1683. Its architecture expresses its seniority within the hierarchy of Old Court: there are two storeys rather than three, although the height is the same as the other three ranges, and it houses communal spaces rather than accommodation, including the Hall and the Fellows' Library.

Old Court is an exquisite small example of the traditional Cambridge court, enclosed on all four sides and open to the sky – rather like an outdoor room. In the days when plague and other diseases were rife, such courts were regarded by some as unhealthy, and the hugger-mugger accommodation may certainly have encouraged infection; indeed, one of the reasons the College was anxious to have access to the other side of the

Left: The inner face of the eastern gate.

Below: A detail from the western gateway arch.

river was 'to enable them to pass to and from the College otherwise than through the town in times of pestilence'. But the seventeenth-century Fellows opted to follow tradition in their new building; and whoever conceived and executed the work, he or they have left their legacy in the untouched splendour that can be appreciated within Old Court itself, from the opposite bank of the Cam, and from King's lawn where, unusually for Cambridge, the outside façade of Clare's South Range appears to be part of King's. Old Court has been called 'a work of art, a piece of history, a practical arrangement of spaces in which to work, sleep, eat and worship – all of these in the same conception'. Paul Mellon, who had rooms on G staircase when he was at Clare in the late 1920s, put it perhaps more succinctly when he called Old Court 'the most beautiful building in England'.

While the overall appearance of the court is its greatest splendour, the details throw light on the original conception and on the lives of those who built and used it. The fan vaulting in the east gateway, by the Porter's Lodge, is, to quote Pevsner, 'the last example of the survival (rather than revival) of this characteristic Perpendicular motif'. The architecture of the east gateway as a whole betrays a number of architectural styles, both 'modern' to the seventeenth century and looking back to the era of Henry VIII, eg the slender two-storeyed semicircular oriel window with its pedimented gable above the inner face of the arch. The windows of many of the rooms in Old Court have high sills, designed to let light in rather than allow the occupants to gaze out – an aid to the serious business of study. Note particularly E staircase, with its fine door and even finer staircase within – a relic of the time when the Master was lodged in rooms here before his own accommodation was built in the north-west corner.

Right: The Chapel, with the Cipriani painting above the altar and the Snetzler organ to the left.

The Chapel

As you enter Clare from Trinity Lane, the College Chapel is the building on the right, but its entrance is tucked into the north-east corner of Old Court itself. The door which gives access to it, with its large shell-hood, is earlier than the current Chapel, dating probably to the late seventeenth century and echoing the shell mouldings over the statue niches on the inner side of the east gateway. Building of the Chapel post-dated that of Old Court by half a century; work started in 1763 and was completed by 1769. Its design combines the enlightened amateurism of Sir James Burrough, Master of Gonville and Caius, with the sophisticated neo-Palladianism of James Essex.

One of the glories of the building is the antechapel which, with its fine octagonal shape lit from above through a glazed cupola, is unique to Cambridge. Another is the painting of the Annunciation over the altar by Giovanni Battista Cipriani, one of the founder members of the Royal Academy; and a third is the rare John Snetzler continuo organ which sits to the left of the altar. It dates to 1755, is still tuned to baroque pitch and is used to accompany motets and anthems. The altar table and the round table in the antechapel are modern pieces by the distinguished furniture designer Luke Hughes.

The two stained glass windows at the east end are Victorian in date, while the two at the west end are early twentieth century. Of these, the one on the north side encapsulates the history of religion at Clare, in depicting its renowned Protestant martyr Hugh Latimer along with Nicholas Ferrar, founder of the Anglican community at Little Gidding which inspired the last of T S Eliot's *Four Quartets*.

SAMUEL BLYTHE MEMORIAL

Samuel Blythe, who entered Clare as an undergraduate in 1652, was elected to the fellowship in 1658 and served as Master from 1678 to 1713, was perhaps the College's most generous early benefactor after Elizabeth de Clare herself. He is remembered principally for his bequest of £6,000 in 1713 for 'the purchase of perpetuall Advowsons and Livings…'. Today, the Blythe Fund represents one-sixth of Clare's total endowment. Samuel Blythe's memory is perpetuated in the annual Blythe feast, and Clare's legacy society is named in his honour.

HUGH LATIMER (c1485–1555)

Elected a Fellow of Clare in 1510, while still an undergraduate, Hugh Latimer was renowned for his blameless life, practical tact and trenchant oratory, and he soon rose to national prominence during Henry VIII's Reformation as a result of his preaching in favour of reform. He became royal chaplain to both Henry VIII and Anne Boleyn in 1534, and Bishop of Worcester in 1535. As one of the king's advisers, he supported the dissolution of the monasteries and his influence on sixteenth-century religious politics was profound. He refused to recant when Queen Mary reintroduced Catholicism and, together with Nicholas Ridley (sometime Bishop of London), he was burned at the stake in Oxford on 16 October 1555. His last words are among the most famous in the English language: 'Be of good comfort, Master Ridley, and play the man; we shall this day light such a candle, by God's grace, in England, as I trust shall never be put out.' His portrait (*below*) now hangs in the Hall.

Below: Eurynome hermits from John Gould's Trochilidae, *Volume 1.*

The Fellows' Library

Situated at the west end of the North Range, the Fellows' Library (not open to the public) was built between 1689 and 1693, replacing a much smaller library above the Chapel. Its position above the kitchens meant that it was neither too dry in summer nor too cold in winter, a factor which proved ideal for the preservation of books and manuscripts – although its site must also have caused frissons of worry among the Fellows about that ever-present medieval danger of a blaze spreading from the kitchens' open fires.

The elegance of the Library's proportions is enhanced by the arrangement of the bookcases in fashionable eighteenth-century country house style round the walls rather than in the more usual

Left: An image from Geoffrey of Monmouth's Historia Regum Britanniae.

Below: The Fellows' Library.

college library fashion, with cases projecting into the room. This layout was in fact reimposed only in the 1950s, after decades when the more usual style was followed; the refurbishment at that time cleared out the build-up of papers and unwanted books that had begun to overwhelm the Library, with the result that – with a few notable exceptions – the present collection includes mainly books published before 1820.

Although the College's early books and manuscripts were lost in the fire of 1521, the Library can now boast a large collection of treasures, most of them acquired through bequests. It is much visited by scholars who wish to consult the rare volumes it contains.

Notable books and manuscripts

The Fellows' Library possesses thirty-five incunabula (books printed before 1500), and about 400 books printed in England before 1640. The earliest are nine theological texts presented in 1557, which bear traces of having been part of a chained library. There are also two copies of the first edition of Newton's *Principia*, one of them heavily annotated in the hand of Newton's contemporary and colleague Charles Morgan, who died as Master of Clare in 1642.

The most notable manuscripts in the possession of the Library are those presented by John Heaver, who died in 1670. They include a thirteenth-century copy of Geoffrey of Monmouth's *Historia Regum Britanniae (above)* and an eleventh-century manuscript of the *Dialogues of St Gregory* which once belonged to King Cnut.

Other treasures are the first translation of the New Testament into Welsh, William Thynne's 1532 edition of the complete works of Chaucer and one of the earliest translations of the Bible into a native North American language (in this case Algonquin). A most generous and striking gift was a remarkable collection of books about birds, presented to his old College in 1954 by the Rev Frank Innes Wane, which includes some outstanding illustrated volumes dating from the eighteenth and nineteenth centuries, several of them published and lavishly illustrated by John Gould (*see previous page*).

THE SILVER

Before the English Civil War the Cambridge colleges, including Clare, were richly endowed with gold and silver plate; but in 1642 King Charles I decreed that all such items must be surrendered to him for melting down to make coinage with which to pay his army. The leader of the men commissioned to collect all the Cambridge plate and deliver it to the king was Barnabas Oley, a Fellow of Clare, who took the risk of keeping back and hiding away five remarkable pieces bequeathed to Clare in 1617 by Dr William Butler, physician to King James I. Along with a few other colleges who also managed to hide some of their treasures, Clare is therefore fortunate in having some of the earliest plate still remaining in Cambridge.

The five Butler pieces include two late sixteenth-century silver 'poison tankards', one incorporating a cabochon of rock crystal, which was popularly supposed to cloud when in proximity to poison, and the other made of serpentine, which was reputed to neutralise all known poisons. Another of his pieces was the Falcon cup, a Flemish wine container and spice box dated 1561, which was later used as a combined tea and sugar canister. But the greatest of his treasures were a unique 23-carat gold chalice and paten made in about 1618 from £260-worth of gold bequeathed for the purpose.

Donations later in the seventeenth century include a magnificent gilt alms dish, given by Samuel Blythe, and candlesticks and a fine pair of flagons, gifts to the Chapel to replace the silver sequestered during the Civil War. Several of the College's fine collection of tankards were given at this time too, as well as a number of spoons.

Much of Clare's silver was donated during the eighteenth century: the collection of tankards provides a good illustration of the evolution in their form at this time, as does the exceptional range of candlesticks owned by the College. Many of the pieces show the influence of refugee Huguenot silversmiths, including a massive rococo parcel-gilt cup and cover by Paul de Lamerie. The College possesses two sets of rosewater ewers and basins, essential requirements at that time for graceful dining. One of them is still in use, as is a Monteith bowl which sometimes makes an appearance in Chapel as a font for infant baptism.

A unique aspect of Clare's silver collection are the 'Greene cups', the result of the bequest made by Richard Greene (1678–1730), Fellow and eccentric philosopher, which he specified must be used for two silver plates or cups, not exceeding a value of £6 each, to be awarded annually to two scholars, one for piety and one for learning. Several of the Greene cups awarded in the twentieth century are examples of excellent modern silver design; and Clare also possesses many fine pieces of antique and modern plate and other treasures donated during recent decades, notably the Nobel Medal for Medicine awarded in 2001 to Dr Tim Hunt, Emeritus Fellow.

THE HALL PORTRAITS

Starting downstairs, left to right, they are:

Richard Terrick (1710–77), student at Clare, admitted 1726; Bishop of London 1764–77; benefactor to the Chapel which he consecrated in 1769.

Charles Townshend (1725–67), Fellow-Commoner, admitted 1747; Chancellor of the Exchequer 1766–7.

Thomas Pelham-Holles, 1st Duke of Newcastle (1693–1768), Fellow-Commoner, admitted 1709; Prime Minister 1754–6 (*below*).

Elizabeth de Clare (?1294–1360), foundress, granddaughter of Edward I.

Thomas Cecil, 1st Earl of Exeter (1542–1623), Fellow-Commoner, admitted 1558, and benefactor.

John Tillotson (1630–94), student at Clare, admitted 1647; Archbishop of Canterbury 1691–4.

Charles, 1st Marquis Cornwallis (1738–1895), student at Clare, admitted 1755; commander of the defeated British troops at Yorktown during the American War of Independence; later Lord Lieutenant of Ireland and Governor-General of Bengal.

Hugh Latimer (c1485–1555), Fellow of Clare, admitted 1510; Bishop of Worcester and Protestant martyr; burnt at the stake in Oxford, 1555.

Sir Henry Thirkill (1886–1971), Tutor 1920–39, Master 1939–58 (*below*).

Upstairs in the gallery (left to right):
Robin Matthews (1927–), Master 1975–93. *Sir Bob Hepple* (1934–), Master 1993–2003.

Outside the Hall above the stairs leading down to the Buttery:
Bronze bust of *Sir David Attenborough* (1926–), student at Clare, admitted 1945; broadcaster and naturalist.

Left: The interior of the Hall, looking west.

Below: The cupola above the Hall.

The Hall

The entrance to the Hall is up a few steps in the middle of the North Range of Old Court, with a handsome cupola *(right)* above it which, like other aspects of this range, shows the influence of Christopher Wren. Inside, on the wall of the lobby outside the Hall facing its doors, is a moulding of the Clare crest which is said to be the only survival of the College's medieval buildings. A fine staircase leads up to the gallery, which retains its delightful late seventeenth-century plasterwork over one of the doors. The Hall's plain wall panelling is of the same late seventeenth-century date, the work of Cornelius Austen who was widely employed in Cambridge at this time, though the swags of fruit and arabesques are a Victorian addition, dating to the early 1870s when Sir Matthew Digby Wyatt was commissioned to refurbish the Hall. The sumptuous plaster ceiling belongs to the same late-Victorian period, and the stained glass depicting the heraldry of College benefactors dates to 1910.

MEMORIES OF LIVING IN OLD COURT

Undergraduates at Clare in the first half of the twentieth century who were allocated rooms in Old Court found themselves living in Jacobean splendour, but sorely without the conveniences of modern life. They were also part of an all-male, rigidly hierarchical society – though the College servants could be as daunting as the dons, as one Edwardian freshman found when he approached the distinguished-looking butler and addressed him as the Master.

Sir Harry Godwin, who came up to Clare to read natural sciences in 1919, recalls being lucky enough to be allocated a share in attic rooms at the top of E Staircase, but rapidly found that rooms in Old Court had 'distinct down-to-earth qualities'. The only heating was an open fire fed from a scuttle on the hearth replenished from a wooden

bin on the landing, to which coal was carried in sackloads by the gyps (household servants): 'We bathed after games in a tin hip-bath before the fire in a canful of hot water brought from the kitchens across the court.' But his bedroom was too far from the fire to benefit from its heat, so on hard winter mornings the water in his hand-basin would be covered by a layer of ice half an inch thick.

In milder weather Godwin could hear 'a rushing cataract of water behind the wall alongside my bed'; when curiosity prompted him to investigate, he found a deep lead-lined wooden trough, which was the gutter draining water from the roof on the King's side of the range to downpipes inside Old Court. He was unsure about the reason for this odd plumbing arrangement: 'whether we needed the water or King's refused to

Opposite: Sir Henry Thirkill, right, and G H A Wilson, left, on Clare Bridge after the opening of Thirkill Court, 1955.

take it, I never discovered.' But it is clearly true that the splendid outer façade of Clare's South Range, overlooking King's lawn, is not interrupted by rainwater goods of any sort.

The only lavatories were to be found in a range of earth closets at the edge of the river, 'privily known as Lady Clare', and at such a distance from some of the rooms that at least one enterprising undergraduate parked his bicycle at the foot of his staircase to make the journey to them shorter. More up-to-date facilities were gradually inserted into the buildings; but Paul Mellon is only one of many who recall strolling across the court in their dressing gowns to take a bath, and those living in lodgings near the College could be spotted in the surrounding streets doing the same.

Undergraduates were expected to attend dinner in Hall on six nights a week and it meant a summons to the Tutor if this rule was broken. Godwin again: 'Food was ample and we were served at the vast oak tables by College servants who, having dispensed the soup or fish, invariably enquired "beef or mutton, sir?" We used the College's heavy Victorian silver at table and for ordinary drinks from the buttery likewise had the use of Georgian or Victorian tankards or silver beakers.' He also recalls College porters carrying meals to students in lodgings 'in large, deep-sided trays, insulated by heavy baize cloths, skilfully borne on their heads'.

Forty years later little had changed, as Duncan Robinson recalls: 'The only baths in Old Court were still those at the foot of A staircase under the Chapel, although gas fires had been fitted into most of the fireplaces to replace coal as the only source of heating in students' rooms.'

It was not until the mid-1960s that the rule that gowns had to be worn in the streets after dark was abandoned. Before that, the Proctor would patrol the streets at night with his two 'bulldogs' – college porters who would chase after anyone suspected of being a gownless undergraduate, and who would enquire, if they managed to catch him, 'Are you a member of this university, sir?' If he was, the fine was 6s 8d (33p in modern money), which would rise to 13s 4d (66p) on a second offence.

Clare Bridge

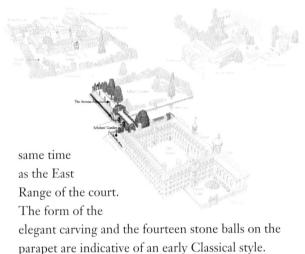

The path through the gateway of the West Range of Old Court leads past the Master's private garden on the right and the Scholars' Garden on the left to the College's bridge over the River Cam. This celebrated bridge is the oldest in Cambridge; work began on it in 1638 very early in Clare's reconstruction programme, not least because it offered a convenient approach from the west for building materials. More importantly, it gave access to Clare's newly acquired land on the opposite bank of the Cam, and marked a considerable expansion in the College's amenities. The mason Thomas Grumbold, father of Robert who was later to play such a central role in the building of Old Court, was paid 3s for a drawing of the bridge, and work commenced in 1638 at the same time as the East Range of the court. The form of the elegant carving and the fourteen stone balls on the parapet are indicative of an early Classical style.

If you stand at the west end of the bridge and look back at the outer façade of the West Range of Old Court – provided the magnificent, 200-year-old copper beech in the Master's Garden allows a

AN UNDERGRADUATE PRANK?

One of the stone balls (the penultimate one on the King's side of the bridge, furthest from Old Court) has a segment carved out of it on the river side. It is not known when or how this occurred. A prosaic explanation puts it down to a fault in the stone; more romantically, it is said to be the result of an undergraduate prank – a bet laid by one College member on another as to how many stone balls the bridge could boast. In order to win his bet, one of them is said to have removed the segment so that one ball could no longer be thus accurately described. Yet another story has it that one of the stonemasons working on the bridge was never paid the final instalment of his fee, so deliberately spoilt part of his work.

view – it becomes clear that the positioning of the gateway is asymmetrical, that it is not in the middle of the range. This is because the North Range, to the left, was designed to be deeper than the other ranges because of the importance of the rooms it contained. There are therefore nine windows to the left of the gateway and seven to the right.

Note the two large yews in the Scholars' Garden, which have become known as Tweedledum and Tweedledee, and which hide a small flight of steps between them. The asymmetry of the west range can perhaps be best appreciated from this garden, and its terrace affords an excellent view both of the bridge itself and of activities on the river.

The fine iron gates at the west end of the bridge are by the same craftsman who made those at the College entrance.

The Gardens

A right turn at the end of the bridge leads through gates and down steps to the Fellows' Garden. Sited on the ancient fenland and bounded on all four sides by water – dykes on three sides and the river on the other – it is, after its redesign in 1947, a fine example of a twentieth-century English garden which, however, retains many of its earlier features.

The building of Memorial Court in the 1920s had deprived the Fellows of their garden on the other side of Queens Road. But they did little with

their new garden by the Cam for several decades – their benign neglect led to Clare acquiring the nickname 'the palace in the jungle' – and the rigours of the Second World War years saw the staff reduced to one ageing gardener. It was therefore a priority after the war to rehabilitate and redesign the garden, and Nevill Willmer, a Fellow and an amateur garden designer of distinction, was entrusted with the work.

Aware of the varied uses to which the garden was put, Willmer's approach was to divide it into

as many different sections as possible, giving each, to quote him, 'a special character with respect to such things as season, colour, scents, privacy and use'. The garden was fortunate in being well protected from winds and enjoyed the high water table afforded by the river; and it was clearly desirable to keep some of the long-standing existing features such as the old brick wall and the mature trees – though the elms had succumbed to disease and had to go. But he also incorporated into his vision his own fascination as a 'Sunday painter' with landscapes.

This led him to create a series of vistas which would be seen from various vantage points and angles. Apparent distances were increased by gradually reducing the width of alleys or by placing brighter orange-red flowers in the foreground and more muted crimson hues in the distance. And perhaps the most magnificent of these vistas saw its expression in the glorious double-sided herbaceous border, planted in blues and yellows and resplendent in summer with delphiniums, which leads up from the riverbank to the sunken garden.

The sunken garden itself represents the more formal element of the design. Enclosed by a clipped yew hedge, with low terraced walls which reach a

Clare College, Cambridge

Opposite: The formal sunken garden in the snow, looking towards the Avenue.

Below right: The sub-tropical garden.

Overleaf: The lawn and herbaceous border looking towards the West Range of Old Court across the river.

flowering peak in May, it has at its centre a pool which depends for its water level on the level of water in the Cam. The two gnarled and ancient apple trees which flank the pool at its north end are relics of the old kitchen garden, and still provide fruit for the College's kitchens; the variety is believed to be Lord Derby. This garden was designed to be used as an open-air theatre, and still hosts college plays in the week after the May Balls in the summer.

Two new yews flanking the north gate enhance the perspective of the view from the causeway to the gate leading out to the 'Dean's Walk', which runs in front of the old kitchen garden wall. This shady walk is planted with white shrubs and herbaceous plants, including a 'pocket handkerchief' tree with its characteristic white bracts, which provide a restful contrast to the bright borders behind. At its west end the path turns sharply south to run along the garden's western boundary in a 'tunnel of gloom' under the overhanging trees. This walk was designed to be deceptive, appearing much longer when viewed from the south end – an effect achieved by progressively varying the width and height of the tunnel and the width of the path.

About halfway along the 'tunnel of gloom', a left turn leads to what was once Willmer's 'scented garden'. This retains many of its original scented shrubs but has been developed into a modern version of the Victorian sub-tropical gardens *(above right)*, which creates its effect through its foliage and shape and the variegation of its planting.

Much of the garden is taken up with a spacious and informal lawn, flanked by a mixed herbaceous border which leads up to the river between two

island beds containing plants of every kind. In spring these are bright with bedding displays, but the reds and oranges of the autumn colour scheme, reflected by neighbouring trees and shrubs, draw the eye to the garden's eastern end at the season's close. The lawn is graced at its west end by Barbara Hepworth's sculpture *Two-forms (Divided Circle)*; and the gate by Wendy Ramshaw that here leads out to the Avenue is a recent (2008) commission, funded from Nevill Willmer's bequest to the College. The fish-eye lens in its central oculus allows the garden to be viewed and appreciated even when it is not open to the public.

The garden can boast a number of fine trees, including an American buckeye chestnut, a swamp

Bottom: Barbara Hepworth's Two-forms (Divided Circle).

cypress and a dawn redwood. This is a 'living fossil' dating back to the age of the dinosaurs. Thought to have been extinct for over five million years, it was rediscovered in a remote Chinese village in 1941 and the Clare example was one of the first to be planted in Britain. Its strong pyramidal silhouette in winter, the fresh spring foliage and good autumn colour provide year-round interest.

Nevill Willmer reflected on his design for the Fellows' Garden in a 1953 article in *Gardening Illustrated*: 'A college garden should essentially be a place in which it is pleasant to wander meditatively

THE GARDEN FLOODED

The advantage to the garden of its proximity to the River Cam is a mixed blessing when the danger of flooding looms. Nevill Willmer recalled the ravages wrought on the yew hedge newly planted in the autumn of 1946 by the vicious winter of early 1947 which froze the ground solid for weeks, followed by the rapid thaw which promptly submerged it in water. And in 1978 the swiftly rising Cam drowned the garden in under twenty minutes (*right*). Brian Arbon, then head gardener, noted that in some places the water rose as much as four feet above the garden, and – to his horror – students were seen punting over his sunken flowerbeds. He almost wondered, quizzically, whether a point was being made from on high, since it was precisely at the

Above: The flooded gardens in 1978.

spot where the river invaded that he had been filmed the previous year for *Songs of Praise*, when he had shown rather more knowledge of gardens than of God. More recently, the gardens flooded again in 2001 and 2002.

against a background of quiet horticultural beauty and, perhaps, to play an occasional game of light-hearted bowls or even spiteful croquet. Moreover, dons being what they are, variety is necessary in the garden, though some resent even this frivolity. The designer must remember that the newly installed Doctor of Science cannot tolerate any red flowers which might detract from his scarlet gown; the Mathematics Tutor has a passion for symmetry among infinitely numerous beds, which is unfortunately not shared by the most recently elected Fellow who, unlike the distinguished octogenarian Senior Honorary Fellow, dislikes any design which has ever been used before. Furthermore, as the plans mature, a perpetual argument ricochets across the high table between the art expert of the college and its elderly but eminent Lecturer in Botany as to whether the garden should enhance the beauty of the college or provide the Botany School with a living herbarium. The latter point of view receives the support of the Medieval Historian, who insists that woad should be grown in the garden.'

AMERICAN CONNECTIONS

Clare's transatlantic connections run deep. It was a sometime Fellow of Clare, Nicholas Ferrar, who assisted Sir Edwin Sandys as governor of the Virginia Company in drafting the constitution under which the first representative assembly convened in America met on 30 July 1619. Over a century later, two Clare contemporaries, Charles Carroll of Annapolis and the Hon Charles Townshend, later Chancellor of the Exchequer, who were both at the College in 1742, found themselves on opposite sides as relations between the mother country and her American colony deteriorated and the American War of Independence began. During that war another Clare man, Charles, 1st Marquis Cornwallis, who was admitted in 1755, presided over the 1781 defeat at Yorktown, when he surrendered with 8,000 men to a combined American and French force under George Washington – the British defeat that was to signal the start of negotiations to end the war that resulted in the Treaty of Paris in 1783. Despite this setback, Cornwallis went on to enjoy a long and distinguished career in both India and Ireland, and his portrait now hangs in the Hall.

The most renowned and influential recent American connection is undoubtedly with the Mellon family, Paul (1907–99) who was at Clare from 1929 to 1931, and his father, Andrew, whose charitable foundations benefited the College and the university. As Paul wrote of his time at Clare, 'Cambridge I loved… its grey walls, its grassy quadrangles, St Mary's bells, its busy, narrow streets full of men in black gowns, King's Chapel and Choir, and candlelight, the coal-fire smell, and walking across the quadrangle in a dressing gown in the rain to take a bath.'

Paul Mellon read history at Cambridge after graduating from Yale. He was the College's greatest twentieth-century benefactor, and also a major

donor to the University of Cambridge, the Fitzwilliam Museum and Clare Hall. He gave generously to the appeal for the Forbes Mellon Library and established the Paul Mellon Clare Trust, to fund in perpetuity major repairs and capital improvements to Old Court, the Chapel, Clare Bridge, the Avenue and the gardens. Only a year after graduating from Clare, he established a student exchange programme between his two *almae matres*, Yale and Clare; the Mellon Fellowships continue to flourish, eight decades later. His generosity is commemorated by an inscription inside the archway of G staircase, Old Court, where he lived during his time in College.

Andrew Mellon, who became the US Ambassador to the Court of St James in 1932, was honoured by the university, which gave him an honorary degree, and by Clare, which conferred on him an honorary Fellowship. His son was to follow in those footsteps, both in 1960 when he became an honorary Fellow of his old College and in 1983, when he returned to Cambridge to receive an honorary degree. At the ceremony the university orator echoed Wren's famous epitaph in St Paul's Cathedral when he said of Paul Mellon, '… si huius munificientiae documenta requiritis circumspicite' (if you require evidence of his munificence, look around you).

35

The New Buildings

After the First World War, Clare's rapid expansion necessitated new building; and it is to the credit of the Fellows of the time that, rather than endanger the architectural integrity of Old Court, they chose to construct their new court on the other side of the Cam and Queens Road, on land which had been acquired in the nineteenth century and was being used for gardens and playing fields.

The Avenue that leads to the new buildings was laid out in 1690 and passes through two sets of fine iron gates, by Warren, of the same period. A relic of the past can be seen outside the gates on the edge of Queens Road, in the form of an old mounting block which Clare men would have used to mount their horses – a contrast with the busy and never-ending streams of traffic that now divide the old from the newer parts of the College.

Above: The exit gates on Queens Road that lead over to the new buildings.

Left: Memorial Court with Henry Moore's
Falling Warrior.

Below: The Forbes Mellon Library.

Memorial, Thirkill and Ashby Courts

The new court was conceived specifically as a memorial to the Clare men who had lost their lives in the First World War. Its architect was Sir Giles Gilbert Scott, who was known for blending his eminent grandfather's Victorian Gothic tradition with twentieth-century modernism. His design moved away from the traditional Cambridge 'staircase' approach to accommodation, which allowed little or no sideways communication, and instead grouped rooms around central landings. This horizontal approach to the design is reflected in the thick band of stone which runs above the first floor windows in a continuous string course.

The monumental memorial arch which forms the entrance to the court is pierced to accommodate a large bell and carries the names of Clare men who died in both twentieth-century world wars. The memorial theme is enhanced further by the sculpture *Falling Warrior* by Henry Moore which has had its home in the court since the early 1960s.

Memorial Court was extended in the 1950s when Thirkill Court was constructed as an extension to its South Range. This new court, named after Clare's distinguished Master from 1939 to 1957, Sir Henry Thirkill, took the issue of

undergraduate comfort further than previously, to the extent that it became usual for overseas students to be allocated rooms here when they first arrived, since its provision of twentieth-century electric heating allowed them to acclimatise gradually to the rigours of Cambridge winters.

When the decision was taken in the 1920s to build a new University Library, and the site chosen was immediately behind Memorial Court, Sir Giles Gilbert Scott was the obvious choice as its architect. His design for the library blended with and respected his vision for Memorial Court, and he conceived the two buildings as linked through a central vista running from the memorial archway through to the library's grand entrance.

THE *DOUBLE HELIX* SCULPTURE

The discovery of the structure of DNA by Francis Crick and James Watson in 1953 ranks as perhaps the most important scientific advance of the twentieth century, laying the foundations for modern genetics. While the breakthrough and those who made it are justly famous, it is less well known that James Watson was at Clare, and living in Memorial Court, at the time. His rooms were on R staircase.

Professor Watson returned to Clare in 2005 to unveil a permanent monument to the discovery: a twelve-foot high representation of the double helix by the sculptor Charles Jencks. Set in the garden at the front of Memorial Court, aligned with the Avenue, visible from Clare Bridge and cast in burnished aluminium, it is an eye-catching landmark, celebrating the College's connection with a scientific milestone.

Inscriptions on the sculpture explain the significance of the double helix and remember

Above: James Watson at the unveiling of the Double Helix *sculpture, 2005.*

the four scientists who made important contributions to the final breakthrough: Francis Crick, James Watson, Rosalind Franklin (whose nephew, Simon, is a Fellow of Clare) and Maurice Wilkins.

It was therefore controversial when, in the 1980s, plans were unveiled to build the College's new Forbes Mellon Library in the middle of Memorial Court, interrupting Scott's vista. Its architect, Clare alumnus Sir Philip Dowson, wanted to make a clear statement about the importance to the College of this new, and much needed, facility. He therefore placed his pavilion boldly in the centre of Memorial Court, but maintained the relationship with the University Library by siting the entrance to the College

CLARE'S NOBEL PRIZES

James Watson – Medicine, 1962
Norman Ramsey – Physics, 1989
Tim Hunt – Medicine, 2001
Mohan Munasinghe UN/IPCC – Peace, 2007

library on the side of the building facing its larger counterpart; the blank wall of solid brick that confronts those approaching from the memorial arch only serves to underline the confidence and assertiveness of Dowson's design. The name of the library commemorates both Mansfield Forbes, whose bequest of his own books formed the nucleus of the undergraduate library in the 1930s, and Paul Mellon, who gave a large personal donation to the library appeal. And the tower of the University Library still looms over both Memorial Court and the bold new College library within it.

The library divides Memorial Court into two, with the halves linked via passages to the left and right of the bulk of the building. A separate court has therefore in effect been created to the west, on the University Library side; this has been named Ashby Court in honour of Eric (later Lord) Ashby, Master from 1958 to 1975.

Lerner Court

The architects of Lerner Court, van Heyningen and Haward, aimed both at a distinctive and contemporary design and to remain sympathetic to Sir Giles Gilbert Scott's architectural vocabulary as expressed in Memorial Court; the new buildings use the same materials and echo its volume and proportions. The aim has been to create both a humane and measured environment and a worthy twenty-first-century addition to Clare's distinguished architectural inheritance.

The court combines two main functions: an accommodation wing and a state-of-the-art study centre facing the University Library. The design reflects both those uses: the accommodation wing has a calm, timber-clad façade while the frontage of the western wing facing the library is of a more imposing brick. The study centre combines an underground, 150-seater auditorium for lectures and films with a spacious, light-filled, double-aspect foyer with a portico and terrace extending out into the court and a double-height music/conference room.

The space created by the new buildings is as important as the buildings themselves. Lerner Court – named for Randolph Lerner, an alumnus and major benefactor of the College – does not simply replicate other courts, but aims to be a more dynamic, landscaped space, and opens up a new north–south pedestrian axis linking the new court with the Forbes Mellon Library and Thirkill Court, and greatly improving access to the latter; it has been laid out by the distinguished landscape architect, Robert Myers.

Above: The study centre in Lerner Court with the tower of the University Library behind.

The Colony

The Colony, Clare's 'third court' situated behind St Giles's Church between Castle Street and Chesterton Lane, has been a central element in students' experiences of Clare for nearly a century, and its origins are nearly as venerable as the College itself.

A parcel of land in the shadow of Castle Mound was first acquired by Clare in 1392. Master Wilflete purchased Castle End Farm in 1466, and a third property was subsequently given to the College by the Master of St Catharine's. By the end of the fifteenth century, the entire site belonged to Clare, and was managed as tenancies.

In 1889, a large part of the land was leased to the Rev Samuel Lewis, whose wife Agnes, with her sister Margaret Gibson, built the majestic Castlebrae house. The equal of many Masters' Lodges, it was seen – as was surely intended – as a statement about the educational emancipation of women, of which the two sisters were ardent advocates.

When Agnes died in 1926, Castlebrae reverted to Clare and was converted into student lodgings. Two further houses, Braeside and Etheldreda, were built, thus formally establishing 'the Colony'. Further accommodation was added over the years through the conversion of houses on Castle Street and Chesterton Lane, and the construction of two new buildings, Castle End (1957) and Castle House (1980s). Castle End was hailed at the time as one of Cambridge's best post-war buildings offering 'freedom in layout and landscaping; freedom from the collegiate shibboleths of enclosed courts; and freedom from the gloom of academic introversion'. Accommodation in the Colony anticipated the growing emancipation

Cambridge students were beginning to demand from the old paternalistic approach of locked college gates and prowling porters; and the addition of Castle House made living or staying in the Colony an active advantage, despite its distance from the College itself.

NB The map above does not give the true position of the Colony, which is situated near Castle Mound, ten minutes' walk from the main College buildings.